*This book is dedicated to ALL mothers,
and especially to my own cherished momma, MLBD.
Thank you for your love, gentleness, and wisdom.
Thank you for listening to my infinite number of math ideas!
Thank you for putting my needs before your own.
Thank you for modeling joy, kindness,
forgiveness, and a positive attitude.
Thank you for sharing your pennies with honorable
organizations that are making our world,
and our children's world, a better place to live.*

Suzy Koontz

*These illustrations are born from my own years of experience
in the honorable profession of motherhood.
They are dedicated to my six children and
my twelve amazing grandchildren*

Jane Woodruff Carroll

A howling, fierce wind rocked the car as Eliza's mother drove her home from gymnastics on a cold April day. Freezing rain beat against the windshield.

But eleven-year-old Eliza barely noticed. She was deep in thought about her friend Freddy Fibonacci. He claimed his allowance would grow so fast he'd be rich in six months. Could it be true? Eliza frowned. She scratched her head. She looked off into space as numbers somersaulted and cartwheeled in her head.

Then she had an idea.

"Mom," she said. "My allowance is four dollars a week. Freddy says that's a constant allowance of 400 pennies a week."

"Yes, I guess that's another way of saying it," said Mom. "But why are you calling it constant?"

"Because it doesn't increase. It stays the same week after week," said Eliza. "I talked to Freddy Fibonacci today. Instead of staying at a constant amount like my allowance, his increases each week because his family uses Fibonacci numbers to determine his allowance.

"The first week, his parents pay him one penny. The second week, they pay him one penny again. The third week, they pay him two pennies. You see, to get the next Fibonacci number, you add together the previous two.

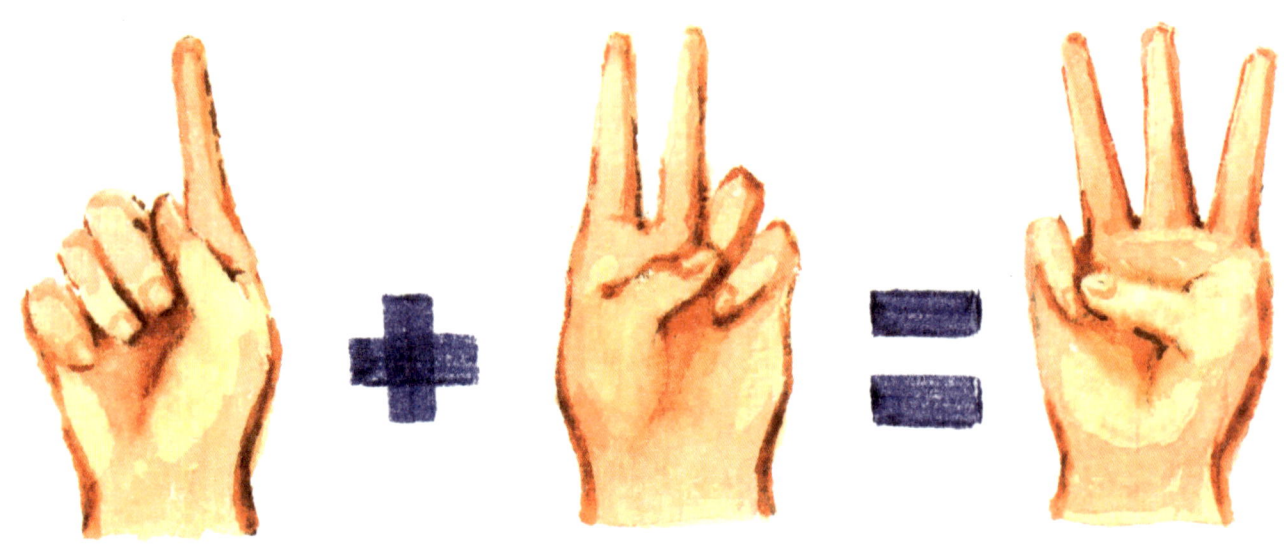

"The fourth week, they pay him three pennies, because one plus two equals three.

"The fifth week, they pay him five pennies, because two plus three equals five.

"And it continues like that, using the same pattern. Mom, do you think we could switch and use Fibonacci numbers for my allowance? The first week, it would only be one penny."

"Are you sure you want to give up your big allowance of 400 pennies?"

Eliza was sure. Her mom agreed to give the Fibonacci allowance a try.

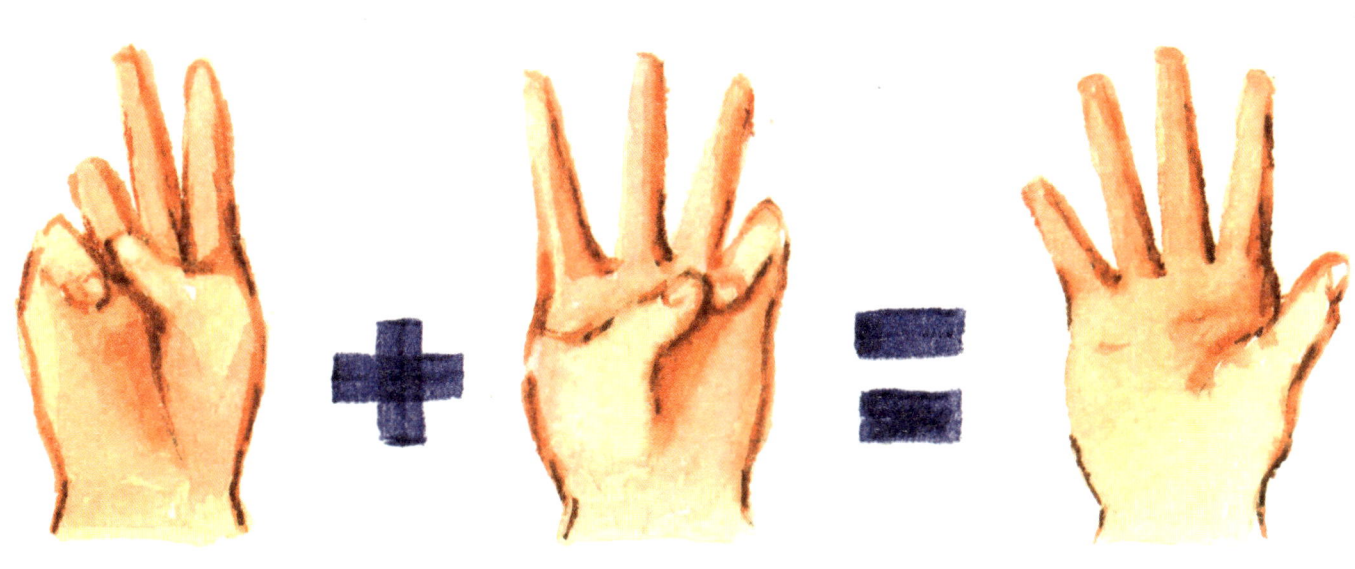

The next Saturday, Eliza found Mom attempting to play a board game with Eliza's younger sister Sarah. While baby Jack chewed the box, Jacob mixed up the money and game pieces, and Rusty the dog wagged his tail.

"Mom, I finished putting my clean clothes away.

My allowance, please. It's one penny this week," said Eliza.

"How about a hug, too," Mom said, "since you're making such a big sacrifice."

"Sure, but it's not really a sacrifice, Mom. Thanks for my allowance," said Eliza as she darted out of the room.

On the second Saturday, Kitty chased a mouse into the kitchen. She jumped on the counter and toppled a half-filled milk glass. Milk and glass flew in all directions. Eliza's mom hollered warnings to stay out of the kitchen while she cleaned up the mess.

From the dining room, Eliza called, "Mom, I finished brushing Rusty's fur. My allowance, please. It's one penny this week."

"Only one penny, again? I'll give it to you later. I'm a little busy right now."

"No problem," said Eliza. She skipped away with a sly grin on her face.

The next week, Eliza found her mom exasperated with Jacob, who had just smeared the family room wall with peanut butter. "I just painted this wall!" sighed Mom.

Eliza patted her arm. "Mom, I polished the door knobs. My allowance, please. It's two pennies this week."

"Only two pennies? Are you sure?"

"I'm sure," Eliza said.

On the fourth Saturday, Eliza found her mom playing the piano with Sarah.

"Mom, I cleaned the rabbit's cage. Phew, that was gross! Are you sure Sarah cleaned it last week?"

"I did!" protested Sarah.

Eliza wiped her brow with the back of her hand. "My allowance, please. I really earned it this week. All three pennies."

"Just three pennies? Okay. But don't spend it all in one place," Mom said with a chuckle.

The next week, Eliza found Mom searching for her car keys.

"Have you seen them?" asked Mom, her hands on her hips.

"They're not in the play kitchen, the refrigerator, or the trash can. I can't find them anywhere."

Eliza thought. "Last night I saw Jacob playing basketball with keys and his boot. Look in his winter boots. Maybe he made a basket."

As Mom shook each boot, Eliza said, "I trimmed Kitty's nails and changed her litter box. May I have my allowance, please? It's five pennies this week."

A boot jingled in Mom's hand. "Thank goodness," she said, as she dumped out the car keys. "I'm sorry, Eliza. What were you saying? Only five pennies?"

"Yes, Mom. Five pennies."

"Eliza, over the last five weeks I have paid you a total of twelve cents," she said. "This Fibonacci allowance does not seem fair to you."

9

"That's called a cumulative sum!" announced Eliza.

"What's a cumulative sum?"

"A cumulative sum of allowances is when you look back and add up all the allowances you've received. The cumulative sum of my Fibonacci allowance so far is twelve pennies. The cumulative sum of my old level allowance of $4 a week would be $20 since five times four equals twenty, or 2,000 pennies. That's a difference of $19.88, or 1,988 pennies," said Eliza.

"That's what I mean! This allowance is not fair. Let's go back to $4 a week."

"It will even out at some point," said Eliza. "Can't we please use Fibonacci allowances for six months? We can stop then if you want to. That's only nineteen more weeks. Please?"

"If you insist," said Mom, looking puzzled.

Nine more weeks of Fibonacci allowances passed. Mom kept urging Eliza to go back to the $4 allowance, but Eliza was determined to continue. Her allowance progressed from 5 pennies to 8, then 13, 21, 34, 55, 89, 144, 233, and 377 pennies.

On the fifteenth Saturday, Eliza found Mom in the midst of mountains of laundry.

"Mom, I helped Dad set a live trap for the mouse."

"She was a big help," called Dad from the other room.

"My allowance, please?" said Eliza. "It's 610 pennies."

"Six hundred ten pennies! That's $6.10. Finally! This is the first week that your Fibonacci allowance is bigger than $4," said Mom. "But you've still been short-changed."

10

"Yes, so far the cumulative sum of my old allowance would be $60, or 6,000 pennies, since fifteen times four equals sixty. The cumulative sum of the Fibonacci allowance is $15.96. That's a difference of $44.04, or 4,404 pennies. But the six months aren't over yet!"

The sixteenth, seventeenth and eighteenth Saturdays passed peacefully for Eliza's family. Eliza's mother paid her allowances of $9.87, $15.97, and $25.84 without complaint.

11

On the nineteenth Saturday, Jacob flushed an orange down the toilet. Eliza found her mom muttering and mopping the bathroom floor.

"Mom, I brushed Rusty's teeth. May I please have my allowance? It's 4,181 pennies this week."

"$41.81. Did you double-check your calculations?"

"Yes, I added last week's allowance and the allowance of two weeks ago to get the allowance for this week. Last week my allowance was $25.84 and the week before that my allowance was $15.97. So 2,584 pennies plus 1,597 pennies equals 4,181 pennies or $41.81. I did the math in my head, and checked myself on paper."

"These allowances are getting expensive!" said Mom. "You must be ahead by now."

"Yes, indeed!" said Eliza. "I'm ahead by $33.45."

"That's a lot of money," said Mom.

12

On the twentieth Saturday, Eliza's mother baked four loaves of bread and two casseroles for the freezer. She left them out to cool. Rusty put his front paws on the counter, and ate one casserole before Eliza's mother caught him. Eliza found her mom scolding the dog.

"Mom, I helped Sarah catch her parakeet. My allowance, please. It's 6,765 pennies this week."

"What? $67.65!" said Mom. "From now on I will call your allowance 'Freddy Fibonacci's awful allowance!' It's costing me far more than I ever expected. How much longer will this go on?"

"You promised to continue for six months. It's been twenty weeks so far. We have four more weeks to go," said Eliza.

On the twenty-fourth Saturday, Eliza found her mom raking leaves in the yard. Eliza hopped up and down with excitement as she said, "Mom, I picked out the burrs from between Rusty's toes. My allowance, please. It's 46,368 pennies this week."

"$463.68? Eliza, you have broken the bank! I'll be so glad to go back to the $4 allowance. How many pennies have you earned over the last six months with your Fibonacci allowances?"

14

"I have earned a grand total of 121,392 pennies, or $1,213.92!" said Eliza.

"What are you going to do with your thousands of pennies?" said her mom.

"I guess I'll put some of it in my college fund," said Eliza, hiding a grin. "Oh, by the way, Dad said he is making dinner tonight."

Upstairs, Eliza quietly made two phone calls.

Later she walked into town with the money jingling in her pocket. Would she find exactly what she wanted? Eliza hunted from shop to shop before she saw it. Perfect!

At home, Eliza slipped into her room, and put the gift on her bed. She couldn't stop looking at it, wondering what Mom would think.

That evening, Eliza's stomach was full of butterflies as she heard Dad say to Mom, "Eliza wants you to see something in her room. Why don't you go upstairs while I fix dinner?"

A few minutes later, she heard her mother gasp, "This is the most beautiful dress I've ever seen!"

Mom looked more sparkly than a million new pennies as she floated down the stairs wearing her swirly new silk dress, a pearl necklace, and matching earrings.

"Oh, Eliza!" she said. "This dress is gorgeous! But isn't it a bit fancy for dinner at home?"

"I have two more surprises," she said. "Look outside!"
There in the driveway waited a long white limousine.
"A limo!" exclaimed Mom. "Where are we going?"

20

"I'm treating you to dinner at Westin Manor!" Eliza said. "We have a reservation for six o'clock."

"And don't worry," added Dad. "We'll save you some cake!"

Mom wrapped Eliza in a big hug. "I guess Freddy Fibonacci's awful allowance has turned into a special evening for my clever girl and me. Thank you, Eliza!"

THE END

A Note to Parents and Teachers

Fibonacci numbers are named after an Italian mathematician, Leonardo da Pisa (1175-1250), who was better known as Fibonacci.

All of us are familiar with the following methods of counting:
1, 2, 3, 4, 5, 6, 7, 8, 9, 10, ... or
2, 4, 6, 8, 10, 12, 14, 16, 18, ...or
5, 10, 15, 20, 25, 30, 35, 40.

Few of us are familiar with the method credited to Fibonacci. The Fibonacci sequence counts numbers in the following way:
0, 1, 1, 2, 3, 5, 8, 13, 21, 34, 55, etc.

The Fibonacci sequence starts with 0, 1. The next number in the sequence is calculated by adding the two previous numbers in the sequence together. For example, 0+1=1. Then 1+1=2, then 1+2=3, then 2+3=5, then 3+5=8, and so on.

Fibonacci numbers are everywhere! In the best-selling book *The Da Vinci Code*, a Fibonacci number sequence is used to unlock a safe. And in nature, flowers often have 3, 5, 8, 13, or 21 petals–all Fibonacci numbers. Look for them elsewhere in nature, as well as in art, architecture, investing, mathematics, and music.

Finally, I highly recommend politely declining your clever son or daughter's request to use Fibonacci numbers to determine their allowance for experiments exceeding 20 weeks. On the other hand, Fibonacci numbers are great for allowances for shorter time periods. And of course, they're great for learning math, too!

Sincerely,

Suzy Koontz

Below is a chart that shows how slowly and then how quickly the Fibonacci sequence accumulates money.

Week	Pennies Eliza Earns each Week	Fibonacci Allowance	$4/week Allowance	Cumulative Sum of Allowances Fibonacci Allowance	Cumulative Sum of Allowances $4/week Allowance
1	1	$0.01	$4.00	$0.01	$4.00
2	1	$0.01	$4.00	$0.02	$8.00
3	2	$0.02	$4.00	$0.04	$12.00
4	3	$0.03	$4.00	$0.07	$16.00
5	5	$0.05	$4.00	$0.12	$20.00
6	8	$0.08	$4.00	$0.20	$24.00
7	13	$0.13	$4.00	$0.33	$28.00
8	21	$0.21	$4.00	$0.54	$32.00
9	34	$0.34	$4.00	$0.88	$36.00
10	55	$0.55	$4.00	$1.43	$40.00
11	89	$0.89	$4.00	$2.32	$44.00
12	144	$1.44	$4.00	$3.76	$48.00
13	233	$2.33	$4.00	$6.09	$52.00
14	377	$3.77	$4.00	$9.86	$56.00
15	610	$6.10	$4.00	$15.96	$60.00
16	987	$9.87	$4.00	$25.83	$64.00
17	1,597	$15.97	$4.00	$41.80	$68.00
18	2,584	$25.84	$4.00	$67.64	$72.00
19	4,181	$41.81	$4.00	$109.45	$76.00
20	6,765	$67.65	$4.00	$177.10	$80.00
21	10,946	$109.46	$4.00	$286.56	$84.00
22	17,711	$177.11	$4.00	$463.67	$88.00
23	28,657	$286.57	$4.00	$750.24	$92.00
24	46,368	$463.68	$4.00	$1,213.92	$96.00

Math & Movement Lesson Plan: Fibonacci Numbers

Grade/Subject: 4th Grade Math
Unit: Operations and Algebraic Thinking - Patterns

Materials:
- Book: *Freddy Fibonacci's Awful Allowance* by Suzy Koontz (ISBN 978-0-9891816-2-4)
- Math & Movement Add/Subtract Mat
- Paper and pencils (one per student)
- Dry erase boards and markers (one for each student pair)
- Chart paper (2 sheets total)

I. Objective
To recognize patterns when adding or subtracting.

II. Common Core Learning Standard
4.OA.5 – Generate a number or shape pattern that follows a given rule. Identify apparent features of the pattern that were not explicit in the rule itself. For example, given the rule "Add 3" and the starting number 1, generate terms in the resulting sequence and observe that the terms appear to alternate between odd and even numbers. Explain informally why the numbers will continue to alternate in this way.

III. Background Information/Prior Knowledge
Students should have a basic understanding of using a rule to solve problems. Students should have a basic understanding of patterns.

IV. Instructional Procedures (70 minutes)
A. Introduction/Motivation (5 minutes)
Set up a dry erase board and marker. Pair students up and have partners sit together around the Math & Movement Add/Subtract Mat. Ask students: "How many of you do chores at home? How many of you get paid or earn an allowance for the chores you do around the house? How many of you would like to earn more money for the chores you do? Today we are going to learn about a girl who was tired of making the same amount of money each week for her chores, so she decided to do a mathematical experiment to see if she could earn more money. First, we are going to review how to use a rule to solve problems."

B. Activity (45 minutes)

- Direct the students' attention to the Math & Movement Add/Subtract Mat.
- Hold up a small white board with "Rule: Begin at 2 and add 5" written on it.
- Ask for a student volunteer to start at number 2 on the mat and then add 5, jumping and landing on the number 7. Place a beanbag on the number 7.
- Have the student jump five more spaces, landing on 12, and then place a beanbag on 12.
- Have the student continue to jump five spaces at a time, and place beanbags on 17, 22, 27, 32, and 37.
- Tell the class to observe the numbers with the beanbags on them and ask if they notice a pattern in the answers.
- Take suggestions from the class. (Answer: The sums form an even/odd number pattern, ending in either 2 or 7.)
- Put the pattern 1, 3, 6, 10, 15 on the board. (Answers: 1+2=3, 3+3=6, 6+4=10, 10+5=15.) Have students work together with their partners to come up with a rule to explain this pattern, using the Math & Movement Add/Subtract Mat and their prior knowledge of patterns. (Answer: Beginning with 1+2=3, take the sum of the numbers (3), add the next consecutive addend in the sequence (3), and continue this pattern.)
- Select a partner pair to come up and use the mat to show how they found the rule to explain this problem.
- Put the pattern 8, 13, 19, 26 on the board (Answers: 8+5=13, 13+6=19, 19+7=26), and have students work together with their partners to come up with a rule to explain this pattern, using the Math & Movement Add/Subtract Mat and their prior knowledge of patterns. (Answer: Beginning with 8+5=13, take the sum of the numbers (13), add the next consecutive addend in the sequence (6), and continue this pattern.)
- Select a partner pair to come up and use the mat to show how they found the rule to explain this problem.
- Introduce the book *Freddy Fibonacci's Awful Allowance*, and explain that because this book contains patterns to solve, the class will stop periodically while reading the book to discuss and solve some problems.
- Read to the bottom of page 3, show the illustrations of Eliza's hands, and write the math sentences 1+1=2, 1+2=3, and 2+3=5. Ask students if they can find a pattern between these three addition sentences. (Answer: Add the previous addend and the sum together to get the next number.)

- Draw the below chart on the board and have the students copy it onto their paper. Make sure there is enough room to continue this chart all the way to the 24th Saturday
- Working together with the students, complete the chart down to the 8th Saturday. Use the Add/Subtract Mat to help with the addition.

1st Saturday	$0.01
2nd Saturday	$0.01
3rd Saturday	$0.02
4th Saturday	$0.03
5th Saturday	
6th Saturday	
7th Saturday	
8th Saturday	

C. Partner Activity (20 minutes)

Have students work with their partners and use the chart to figure out how much Eliza will receive on the 24th Saturday, and the total amount of money she will have at the end of the 24 weeks. Ask students to show their work. Assist students as needed while circulating around the room. When the students are finished, ask them to write an explanation of the rule they used to discover the pattern, and how they arrived at their final answers.

V. Assessment

Assess students informally by monitoring and observing them while they are completing the partner activity. Assess students formally by having the students create a sequence of numbers that forms an addition pattern, and the addition problems needed to produce that pattern. Have the students also provide an explanation of the rule they used to create the pattern and demonstrate why their answers are correct. Next, provide a teacher-created sequence of numbers that forms a pattern. Have the students write down the addition problems that produced this pattern and the rule that explains this pattern.

VI. Closure (20 minutes)

When students are finished with their charts, have them gather back at the meeting area. Finish reading the book. Ask for a volunteer to summarize today's math lesson. Ask the class "Why is it important that we look for patterns when solving math problems? Can you think of any situations outside the classroom where we might need to look for patterns to solve a problem? If you were Eliza, what good things would you do with your large allowance?"